ROYAL
OBSERVATORY
GREENWICH

T0136903

Stars

Dr Greg Brown

Royal Observatory Greenwich
Illuminates

First published in 2021 by the National Maritime Museum, Park Row, Greenwich, London SE10 9NF

ISBN: 978-1-906367-81-7

At the heart of the UNESCO World Heritage Site of Maritime Greenwich are the four world-class attractions of Royal Museums Greenwich – the National Maritime Museum, the Royal Observatory, the Queen's House and Cutty Sark.

rmg.co.uk

A CIP catalogue record for this book is available from the British Library.

Typesetting by ePub KNOWHOW
Cover design by Ocky Murray
Diagrams by Dave Saunders
Printed and bound by
CPI Group (UK) Ltd, Croydon, CR0 4YY

10 9 8 7 6 5 4 3 2 1

About the Author

Dr Greg Brown is an astronomer working at the Royal Observatory Greenwich. In his time in research at the University of Warwick, he studied some of the largest explosions in the Universe and the supermassive black holes hiding in distant galaxies. Combining a love of science, comedy and acting, Greg moved into science communication, where he has been eliciting anguished groans from his audiences ever since.

Entrance of the Royal Observatory, Greenwich, about 1860.

About the Royal Observatory Greenwich

The historic Royal Observatory has stood atop Greenwich Hill since 1675, and documents over 800 years of astronomical observation and timekeeping. It is truly the home of space and time, with the world-famous Greenwich Meridian Line, awe-inspiring astronomy and the Peter Harrison Planetarium. The Royal Observatory is the perfect place to explore the Universe with the help of our very own team of astronomers. Find out more about the site, book a planetarium show, or join one of our workshops or courses online at rmg.co.uk.

Contents

Introduction

Ten billion trillion.

Oh...sorry...big number warning! That probably should have come before...

Ten billion trillion, or a one with twenty-two zeros after it, is a rough estimate of how many stars there are in the entire observable Universe. It's a little more than there are grains of sand on all the beaches of Earth – which while an interesting piece of trivia, has never struck me as a particularly good way of trying to imagine such an impossibly large number.

It is only an estimate, of course. No one has quite gotten around to counting them to check. Instead we take an educated guess. Stars cluster together in vast complexes, called galaxies. Determine the average size of a galaxy (somewhere between 100 million and 100 billion stars) and a rough count of the number of galaxies in the Universe (somewhere between a few hundred billion and 10 trillion), multiply them together and you've got your estimate.

Then of course, choose just one number out of the wide range of possible values you've been left with to quote as the first line of your book. It's more effective that way.

Humankind has studied the stars we see for generations. We drew patterns in them, called constellations, to fit our myths, and later to make discussing individual stars simpler. We've mapped them for the purposes of navigation and used their apparent near-perfect clockwork motions for timekeeping.

But more than anything, perhaps, we've wondered what they are. It's taken thousands of years and some of the greatest minds in science to come from ancient legends involving the spirits of the dead or the light of the gods to a less fantastical but arguably far more impressive conclusion. That they are almost unimaginably large spheres of super-heated gas, radiating light and heat out into the cosmos and powered by the process that turns the simplest materials in our Universe into the stuff that makes up everything around us.

And it is this that we'll discuss here. I'll show you how stars go from a vast cloud of mostly empty space to the luminous beacons we see today. I'll show you how mass really does matter when it comes to stars and take you on to the inevitable ends of their exceptionally long lives. You'll learn how astronomers study these stars and see how much there is yet to understand. Finally, at the end of the book, there is a glossary of terms for words written in **bold font** to cover the more complicated terminology.

So.

Where to start?

Our Sun: A Model Star?

It's probably no surprise to you, nor to towel-wielding hitchhikers of the galaxy, that space is big. This causes a few problems for astronomers. Not only are most objects out of reach of even our fastest spacecraft, but that distance also makes them appear very faint. Their light is spread out over such a great area by the time it reaches our eyes that the vast majority are completely invisible without aid.

Telescopes can help with this. Designed effectively as vast eyes capable of gathering far more light in one go than our much smaller eyes can, they can peer into the distance and reveal objects far fainter than we could see on our own.

The Royal Observatory Greenwich is home to the Great Equatorial Telescope, also known as the '28-inch' (71 centimetres) for the width of its **aperture**. Even 130 years after its production, it remains a sizeable **refracting telescope**. At the time of writing, however, the largest telescope still under construction is the European Extremely Large Telescope (EELT). It will have an aperture of 39 metres across, enabling astronomers to see objects at least 15 million times fainter than the human eye can. Astronomers also make use of special cameras to make their observations. By storing up light in a long exposure, they can see even fainter objects than a 'live' view can.

And yet, as powerful as telescopes and cameras can be, distance still limits us in our study of stars. What would be great is if we had one really close by that we could study in more detail. We could then apply what we learn about it to other stars. In science, this is called finding an analogue – an object that shares features with other objects, but is much easier to study, perhaps because it is particularly bright or particularly close.

Of course, as luck would have it, we have just that! The Sun is our very own star. Big, bright and close-by – perfect for study. But before we get carried away, we have to answer one question: is the Sun a *useful* analogue? In order to be useful, an analogue must be at least somewhat representative of the other objects we are trying to study. In medicine, a new treatment can be tested on a sample of people to check that it is both effective and safe. However, if the medical trial is conducted only on males between the ages of 25 and 35 each with at best a mediocre exercise regimen, the treatment might be able to be shown to be safe for the author of this book, but it would hardly be representative of the public at large.

So, how does the Sun shape up in the grand scheme of the Universe? Does it look like other stars, or does it represent just a small slice of the whole? To answer this, we'll need to look at the vital statistics of the Sun and compare them to other stars – a cosmic top trumps.

Luminosity

We begin with quite possibly the simplest feature of an object in space – how bright it is. And yet we've already hit a problem. As mentioned before, the further an object is from Earth the fainter it will appear to be. A star produces more light than a desk lamp[1], and yet, at night, we know

---

[1] Citation needed.

which will be useful to read by and which will not. So, to know how much light an object is actually putting out, called its **luminosity**, we also need to know how far away it is.

Determining distances in space is not easy. There is no tape measure long enough to reach from here to the North Star, Polaris. Even if there were, we wouldn't have a ship fast enough to get the other end there within our lifetimes – or the lifetimes of our great-great-great-great grandchildren, for that matter.

But there is hope. Astronomers use something known as the cosmic distance ladder. No single way of measuring distances works across the entire Universe. This is the same as on Earth – some methods work well in some situations and fail utterly in others. Using a ruler to measure the Grand Canyon would be just as foolish as trying to determine your inside leg measurement with GPS. Instead, astronomers use one method to determine distances to nearby objects, then use what they learnt there to inform a method useful for somewhat more distant objects and so on. Each 'rung' of the ladder is dependent on all of the rungs below it.

At the bottom of the ladder we have radar ranging – literally bouncing a pulse of radio waves off another object and waiting for the blip to return. A space sonar, if you will! Even with the vast speed of light, this method is only useful within our own solar system, but it can help us determine things like the width of the Earth's orbit around the Sun.

Next comes **parallax**, a method the vast majority of us use daily without even realising it. With two eyes comes two slightly different perspectives on the world. Try raising your thumb at arm's length and then close one eye, then the other and back again. Notice how your thumb appears to shift against the more distant background. Because your brain knows how far apart your eyes are, it can measure the shift of your thumb as an angle. Then, without any input from you, it does some simple trigonometry and determines the distance to your thumb[2]. In exactly the same way, astronomers can use the fact that, over the course of 6 months, our planet Earth moves almost 300 million kilometres in space in its orbit around the Sun. This huge baseline can act like the distance between our eyes, the obvious downside being that instead of being able to take both images at the same time as our eyes do, astronomers must wait six months between them. A background of distant stars and

---

[2] Perhaps unfairly, suggesting to your maths teacher that you don't need to take a trigonometry exam because your brain does it all the time anyway does not tend to have the desired outcome...

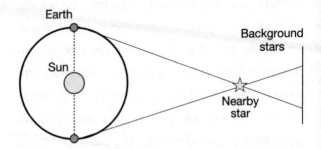

Parallax forms the second rung on the **Cosmic Distance Ladder** – the movement of the Earth, plus a little geometry, allows astronomers to determine the distances to some nearby stars.

galaxies can then be used to measure the apparent shift of nearby stars.

However, even with such an incredibly large baseline to use, parallax can only take us so far. Angles in astronomy are typically measured in degrees with 360 degrees in one full circle. Each degree can be split into 60 arcminutes, and each arcminute split again into 60 arcseconds. This is an incredibly small angle with 1.3 million of them fitting inside a full circle. And yet, the parallax shift of the closest star to the Sun, Proxima Centauri, is less than even this. While modern-day instruments, like those on the European Space Agency's *Gaia* spacecraft, can measure parallaxes down to a tiny fraction of an arcsecond, many of the stars in our galaxy are simply too distant to be measured this way.

And now, the thing that has been making our life so difficult comes to the rescue. Distance makes objects appear fainter, meaning they are harder to observe. But that also means if you happen to already know how bright something *should* be, and then you measure how bright it *appears* to be, you can calculate how far away it *must* be. For this method to work, you need a **standard candle**, an object (often, but not always, a type of star) that once identified is always the same brightness no matter where it is found in the Universe. A number of standard candles exist. Some stars pulse, getting regularly brighter and fainter like a lighthouse at night, and some of these **variable stars** pulse faster if they are fainter. Identify one of these, measure how fast it pulses, and you know how bright it must be. Some **supernovae**, explosions related to the deaths of stars, are always the same brightness no matter where they occur. Each can be used to help extend our cosmic distance ladder well beyond our own galaxy and out into the depths of the Universe.

With the distance to a particular star determined, it is easy[3] to find its luminosity, and so our first vital statistic is done. Our Sun, it turns out, puts out about 4×10^{26} watts, or about 10 million billion billion lightbulbs (so just a bit brighter than a desk lamp then). One thing to learn about astronomers is that they are inherently lazy. Whenever numbers get big and unwieldy like this, they simplify things. So we scale things so that the

---

[3] There is nothing easy about this.

luminosity of the Sun is simply one **solar luminosity** or $1L_\odot$. Of course, some bright spark told Google this, so now if you search 'luminosity of the Sun' chances are you get back $1L_\odot$, which is about as useful as looking up the number of calories in a sandwich and getting back the answer 'one sandwich's worth'. Technically correct – but not exactly useful...

So how does the Sun compare to other stars? Well some stars have luminosities only a few times larger or smaller than the Sun – but many are rather more extreme. The faintest stars can be 100,000 times fainter. The brightest – over a million times brighter. So does the Sun represent other stars well in luminosity. Well...no, not really.

Temperature

It's probably not stretching the imagination to suggest that the Sun is kind of warm.

A bit more than just tepid. Toasty, perhaps. But short of shoving a thermometer in the thing[4] how can we tell just how much of a fever the Sun, or indeed any star, really has?

Strangely enough, this is a task that you can attempt with just your own eyes. Go out at night to a dark site, wait for your eyes to adjust and then stare at the stars[5]. In time, you'll notice that they don't all look the same. Not only are some faint while others are bright, you might just be able to tell that they have different colours. The bright right shoulder of Orion the Hunter is a deep orange star called Betelgeuse, while his left knee is a blue-white star called Rigel. From just this observation alone, you now know that

---

[4] Any volunteers?
[5] The Royal Observatory Greenwich wishes me to suggest it is unwise to do the same with the Sun...for some reason...

Betelgeuse is a cooler star than Rigel, and by some margin.

This is because the colours of light are not random. They are due to the physical properties of whatever the light is emitted by or interacts with. These hues together form a rainbow or **spectrum** of light stretching from blue at one end to red at the other. In reality, this is only what is known as the **visible spectrum** of light, colours our own eyes can see. In truth the **electromagnetic spectrum** extends well beyond this – past blue into ultraviolet, X-rays and gamma-rays and past red into infrared, microwaves and radio waves. Every shade has a certain energy associated with it, so a **photon** of blue light has a greater energy than a photon of green light, which in turn has more energy than a photon of red light. The hotter an object is, the more energy it has available to radiate as light, and more of that energy goes into higher energy types of light. That is, the hotter an object is the bluer it will appear.

It is worth saying here that 'bluer' can simply be astronomer shorthand for 'further down the blue end of the spectrum', which itself means 'higher energy light'. So in the same way that blue light is 'bluer' than red light, ultraviolet light is bluer than blue light. Similarly, 'redder' means lower energy, so infrared light is 'redder' than red light.

We can see this in our everyday lives when we heat objects up to high temperatures[6]. Objects are uncomfortably hot, and even dangerously so, long before they begin to produce visible light. At this point they are radiating light, just light you can't see called infrared. Get them hot enough though and they can glow 'red-hot'. Keep going and you may reach 'white-hot', where all the colours

---

[6] Assuming we all work in steel mills.

of light combine together with no huge bias towards any particular colour (at least to our eyes). This progression is seen in celestial objects as well. Planets and moons are all usually very cold, and so only produce infrared light (heat), though we can see them due to them reflecting the visible light from other sources. Cool stars are red, middling stars are yellow or white, while the hottest stars have a strong blue tinge to them.

How cool is cool? And how hot is the hottest? Well if you measure the colours of light coming from a star using a **spectroscope**, a device used to split out the colours of light, you can plot the brightness of a star in each colour and produce its spectrum. The shape of the spectrum, and where it peaks, tells you its temperature. Our own Sun, a yellow dwarf star, is somewhere in the middle with a surface temperature of 5,800 **kelvin** (5,500°C). 'Cool' red stars still swelter to the temperature of 3,000 kelvin (2,700°C), while the hottest stars can exceed 20,000 kelvin (19,700°C). A rare type of star called Wolf-Rayet can get up to 200,000 kelvin (honestly, at this point, degrees Celsius and kelvin are basically the same).

So, does the Sun accurately represent all of the temperatures of stars in our galaxy?... Nope.

Mass

Weighing stars is not an easy task, and it turns out asking them directly is a massive[7] social faux pas. Navigating one onto a set of scales is equally difficult, so we must turn to the most useful indicator of mass we have – gravity.

All things that have mass have gravity. If we were to be more precise and unnecessarily wordy, we would say that

---

[7] As you might imagine, this joke took me all day.

all things that have mass distort the fabric of spacetime, producing a curvature that causes objects to follow worldlines that also curve, looking to an external observer like an attractive force that we would refer to as gravity. But that would be way too complicated – and the subject for an entirely different book. For now, all that matters is that stars produce strong gravitational pulls that cause planets and other solar system objects to orbit them.

But sometimes stars are not alone. Pairs, trios and even higher numbers of stars can be found orbiting one another. Thanks to Newton's Third Law of Motion[8], every action has an equal and opposite reaction. This means that when a planet orbits a star, the star also orbits the planet. To be more accurate, we would say that they both orbit their combined centre of mass

---

[8] To be clear, he discovered it. He didn't force* it on us all.

*Another day well spent.

or **barycentre**, a place between the two objects that is like the balancing point on a see-saw or lever. For a planet and star, the huge difference in their masses means that their barycentre is often inside, though not quite at the centre of, the star. Just like an unevenly loaded lever, where you have to move the handle a long way to make the heavy load move just a little, the light planet is flung around on a wide orbit, while the heavy star merely wobbles a bit. For two stars orbiting one another, while rarely exact matches, their more equal masses mean their barycentre is closer to being equidistant between the two. Each star moves noticeably around this point and this **binary** dance can be seen using telescopes from Earth.

The Great Equatorial Telescope at the ROG was one such telescope used to observe these binaries. Astronomers observed many suspected binary star systems around the turn of the 20th century, and tried to determine their orbital period – the time it takes for the stars to go around each other once. However, this was not a quick task to complete. The shortest period binaries are close together, so close that from the distance of Earth they appear as a single point of light in the sky, even when observed with a powerful telescope. In order to be far enough apart to be seen as two separate stars, they would have orbital periods of hundreds, thousands, even hundreds of thousands of years. The closest binary to Earth includes the very closest star: Proxima Centauri. It is a tiny red dwarf star orbiting around a further pair of stars called Alpha Centauri AB. While A and B orbit each other once every 80 years, Proxima orbits the pair once every 550,000 years. Only careful observation over a long time can show the slight shifts in position that has led to us realising these stars orbit one another.

Modern instruments are far more precise and so can reduce the difficulty in observing these pairs. We can also use the Doppler effect to help us further. When a car passes you on the road, the sound of its engine seems to change. On its way towards you, the sound is higher pitched than after it passes you and leads to the well-known 'NEEEooooowww' effect[9]. This effect is due to the bunching of sound waves when the car is approaching and the stretching of them as it recedes.

The same happens for light. An object moving towards you has its light bunched up slightly, shifting its colour towards the blue end of the spectrum – known as **blueshift**. When travelling away from you, the object appears redder than it should be as its light is stretched – called **redshift**. Modern day instruments can measure

---

[9] A technical term, if ever there was one.

tiny shifts in the colours of light and thus determine whether the object is moving towards or away from you and how fast. Watching this velocity change as a binary pair orbits each other can be another way to determine their period, and also obtain the speed of their orbits. Thanks to the work of scientists like Johannes Kepler and Sir Isaac Newton, we can use the period of the stars and their speeds to determine their masses by understanding that all of this motion is due to gravity.

So what do we learn?

Well first off it is important to note that these direct observations come from binary stars alone. Bias is a big issue in science. Not personal biases in this case, though a scientist needs to be aware of them to avoid that bias spilling over into their experiment. Instead, some of the main biases that cause problems in science are observational and methodological – that is, biases that come from how you design your experiment and how easy it is to get an answer back. In **exoplanet** science, many of the planets we have discovered are big and close to their stars. This is partly because these types of planets are common, but it is also true that our experiments are geared towards finding those types of planets preferentially. Big planets block more of their star's light, pull harder on their star to make it wobble and are simply easier to see! Close in planets take less time to orbit, so they block their star's light more often, or make it wobble more quickly. The point is we should always be on the look-out for what effect the methods we have used have on their outcomes.

Nonetheless, the results are clear. Stars come in all sorts of sizes. The smallest are just less than a tenth of our own Sun's mass. The largest that we know of are over 150 times the mass of the Sun. And while it's unlikely many such stars still exist, some models of how stars form suggest they may get far, far bigger under ideal circumstances.

So – is the Sun a good analogue for the masses of stars? No. Definitely not.

So, a Failed Analogue?

Things do not look good. Stars range in brightness across 11 orders of magnitude – that is, the faintest star is 100,000,000,000 times fainter than the brightest. While all 'normal' stars are hot, they range from merely very hot (2,700°C) to ridiculously so (200,000°C). And even simply amongst the stars we know of, they can weigh-in at an ultralight 0.08 **solar masses** and all the way up to a gargantuan 150 solar masses. Our moderately bright, moderately massive, merely 'somewhat balmy' Sun represents just a tiny fraction of the whole.

So is the Sun a useless analogue?

Well, no actually! Yes, it's true that the Sun is only truly representative of stars that are just like it, but that doesn't mean we can't use what we learn from it and apply it to stars that are quite different. As long as we bear in mind that it's possible that much larger or smaller stars, for example, may behave somewhat differently, there are still some things that may be similar. Stars are mostly made of the same stuff, they begin their lives in very similar ways and much of their middle-aged life shares similarities too. The same main process that fuels our Sun fuels other stars as well, and using the Sun as a laboratory for the physics of how materials act at high temperatures and in high pressures gives us clues to how other stars behave too. We can also use our understanding of physics generally to take a guess at what differences may be seen in stars that are not well represented by the Sun.

So what exactly do we learn from the Sun?

Appearing big and bright, if not actually big and bright as we've just learned, the Sun has been observed scientifically for centuries. While no doubt the odd early astronomer part-blinded themselves in the attempt, the clever amongst them learned to project an image of the Sun onto a surface that could then be viewed far more safely. Later they started to use **filters** that dimmed the Sun's light down considerably, allowing astronomers to use telescopes to observe in more detail. Some filters would even focus on specific colours of light produced by particular chemical elements, like hydrogen, allowing us to focus on specific parts of the Sun. They could even block out the overpowering light coming from the surface with a **coronagraph**, or just wait for a solar eclipse, allowing them to observe the much fainter outer atmosphere of the Sun.

What they discovered was incredible. Much of the light we see is emitted by a surface known as the **photosphere**, though this is no solid surface like the crust of the Earth. It churns constantly as hot bubbles of gas rise, cool, and sink, like the water in a boiling kettle. These chunks of superheated gas can be seen changing in real time as **granules** that pepper the surface of the Sun. In reality, we see not a physical boundary of the Sun, but a point where the material of the Sun becomes opaque to our eyes. It's like peering into the depths of the ocean. The water can be almost crystal clear, but inevitably light will only travel a certain distance before it is blocked by the combined opacity of the water it has travelled through. Similarly, we can peer into the atmosphere of the Sun a certain distance before the gases become dense enough to block our sight. It is this that we see as the surface.

On top of its bubbling granules, we have bright **plages** and dark **sunspots** – regions that are unusually hot or cold, though often by only a few hundred degrees Celsius. Sunspots in particular have fascinated solar astronomers. They come and go in a pattern that repeats both in where they are found on the surface and in how many appear at any one time. Husband and wife team, Edward and Annie Maunder, were experts in solar astronomy and photography working

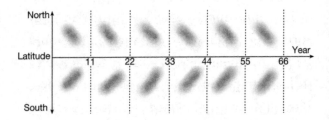

The **Butterfly Diagram**, first used by Edward Walter Maunder, is a record of where and how many sunspots are on the surface of the Sun at any one time. The pattern the sunspots make, repeating every 11 years with the solar cycle, has led to the name of the diagram.

at the Royal Observatory at the turn of the 20th century. They popularised the 'butterfly diagram', which records where and when on the Sun they appear.

A clear 11-year cycle is apparent in our Sun's spots, one that is now known to come from how the powerful magnetic field of our Sun changes over time – changing shape, strength and even completely flipping its north and south poles once every cycle. The sunspots occur where this field bursts out of the surface and up high above the atmosphere, cooling the gas around it.

The atmosphere itself is a mystery of sorts. Held high above the surface, it should be much cooler, and yet has a temperature well in excess of a million degrees Celsius. How this gas is heated remains unclear, though the Sun's powerful magnetic field is expected to play a role.

Magnetism comes in again when the Sun's surface erupts in **coronal mass**

ejections. Billions of tonnes of material are flung off of the Sun at speeds of up to 2,000 miles per second, accelerated by the rubber band-like snap of lines of magnetic energy rebounding back into place.

In short, our peaceful, life-giving Sun is far from the tranquil, eternal ball it seems to be. We may not be able to see far inside of our Sun, but by comparing what simulations and models of the Sun built on basic physics predict to what the Sun actually looks like from the outside, we can take an educated guess. Then, by changing the properties of our simulated Sun to those of other stars and comparing them with our limited observations of them we get closer to understanding them all.

In any case, the Sun is the only star in our solar system, the only star close enough to study in this manner. It's not like we have much choice! It may not be a perfect example of every star in the sky, but it is a great place to start and many of the discoveries we have made about stars have been sparked by the study of the one we have in our own solar system.

So it's lucky we had one, I guess!

The Nuclear Furnace

One of the most important questions about the Sun is perhaps the most obvious: what powers it?

We know now that the amount of energy coming out of a system must equal the amount of energy going into it. Your phone can only provide as much use as you've given its battery charge overnight – too often less than you intended. When attempting a marathon, besides making sure your affairs are in order, it's important to provide your body plenty of energy in part to achieve a good time, but mostly to ensure that you didn't need to get your affairs in order after all. This all seems obvious enough now, but there was a time when this was less clear in the eyes of scientists.

While many scientists and philosophers[10] had made the suggestion that nothing is ever truly lost, it wasn't until the mid-1800s that the concept of the conservation of energy was generally accepted. This states that energy cannot be created or destroyed – it can only be converted from one form to another.

The chemical energy in coal or gas, the **kinetic energy** of wind, or the radiation from the Sun is converted into electrical energy in power plants/turbines/solar panels, which is then stored as chemical energy in your phone's battery to be released as electrical energy again. The chemical energy of your pre-marathon banana is converted into the kinetic energy

---

[10] And the occasional singer-songwriter.

needed to finish the race/avoid coming to your untimely end. In all of these processes, some energy is wasted, as heat, sound, light and so on[11], but if you add it all up you find the total amount of energy never changed.

So, now the issue. Scientists at that time had long since known how far away the Sun was and had realised it must be producing a vast quantity of energy. Only now, rather than have this energy just spontaneously be produced in the Sun, it had to be released from somewhere – converted from one type of energy to the light energy they could see and the many other forms of light they had yet to discover.

Some, physicists Lord Kelvin and Hermann von Helmholtz for example, suggested that the Sun could be powering

---

[11] Hence the common sight of the bioluminescent long-distance runner.

itself by slowly collapsing. Dropping a ball from a great height on Earth causes it to accelerate downwards rapidly[12]. This is converting **gravitational potential energy** into kinetic energy. Drop enough balls simultaneously onto an object with enough gravity and the ensuing cramming, compression and friction would produce vast amounts of light and heat. Imagine now that you are progressively dropping the outer layers of the Sun onto its own core and you have a rough idea of what Kelvin and Helmholtz were suggesting.

There were only a few problems with this. First it would require that the Sun were shrinking – a somewhat alarming prospect. However, at the time the rate at which the Sun would have to be shrinking would not have been measurable. The

---

[12] If, in your own test, the ball instead flew upwards, you were holding it upside down. If the ball flew sideways, it was actually a bird. Or perhaps a plane.

other problem was that this method doesn't last very long – at least not where astronomy is concerned. At the time the Earth, and by extension the solar system, was thought to be many millions of years old, though how many millions was up for debate. Kelvin himself believed it was a few tens at most. By the 1910s this increased to over 1 billion years thanks to the use of **radiometric dating,** a method that measures the abundance of different radioactive elements and their products to determine how long an object has been around. Now we know the Sun to be about 4.5 billion years old. The contraction of the Sun could sustain it for at most a few tens of millions of years – fine for Kelvin's view at the time, not so good now we know better.

So what could it be?

Thanks to the work of scientists like Cecilia Payne-Gaposchkin, astronomers knew that the Sun was made mostly of hydrogen and helium – indeed helium was first discovered in the Sun, hence its name based on the sun-god Helios. While hydrogen might seem like a great source of energy[13], without oxygen to supply a chemical reaction, it cannot burn – and it would be woefully insufficient to power the Sun even if it did.

To the rescue came the last piece of the conservation of energy puzzle and with it the most famous equation in science: $E = mc^2$. It was the brainchild of one Albert Einstein, scientist and anti-frizz shampoo and conditioner enthusiast. In it, the energy, E, of an object can be expressed as its mass, m, times the speed of light, c, squared. In reality it's a bit more complicated than that[14], but in essence mass and energy are the same thing. One can be converted into the other.

---

[13] 'No smoking near the hydrogen tanks, sir.'

[14] Did I say astronomy was complicated – just wait till you try relativity.

What's more, because the speed of light is a big number, and the square of it is a really big number, just a little bit of mass converts into a vast quantity of energy. The average UK household uses about 16,000 kilowatt-hours (5.8×10^{10} Joules) of energy in electricity and heating each year. To power that, purely through the perfect conversion of mass into energy, you would need just 0.00064 grams of matter – or about the mass of two fully dried adult fruit-flies[15]. Of course, that's if you could somehow spontaneously make matter convert itself to energy, which is... difficult. The Sun, however, has a way.

We interrupt this science text for a culinary interlude. A recipe for the energy output of the Sun – serves eight[16]. Take four **protons**, one of the simplest particles in the Universe. Fairly heavy and

---

[15] Best not to ask how I know that.
[16] Nine, if you count Pluto.

with a positive charge, it is the central point, the nucleus, of the simplest atom in the Universe: hydrogen. Place in a stupendously high pressure furnace at gas mark 1,079,128 (15 million degrees Celsius) and allow to boil vigorously. The resulting flurry of activity will cause protons to collide with one another, occasionally sticking together and falling apart again. Rarely, these stuck proton pairs will change, one proton converting itself into a **neutron**. More chance collisions and a long simmer later, and four particles will have stuck themselves together. The result – a nucleus containing two protons and two neutrons: helium. Serve with a **plasma** jus.

This subatomic cookery is known as nuclear fusion – specifically the proton-proton chain, the main form of fusion in our own Sun.

But here's the interesting bit. Add up the masses of the four protons that went

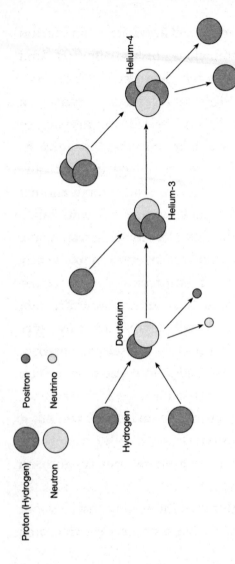

The recipe for energy in the Sun. In total, four hydrogen nuclei (protons) fuse together to become the larger element, helium, producing energy, neutrinos, and positrons (the antimatter version of an electron) as by-products. It may look complicated but it's simpler than a recipe that measures in 'cups'.

44

in and compare it with the combined mass of the helium nucleus that came out and the other bits that flew off in the process. You'll find you have less stuff than you started with. Not a lot less, a drop of about 0.7%. Where did this mass go? Into energy, of course, as Einstein predicted. One single reaction produces about enough energy to… well… not really do very much at all really. Lift 1 trillionth of an apple 1 metre off the ground? Or equivalently lift an apple 1 trillionth of a metre off the ground. Hardly stellar. But the Sun is doing this one reaction 10^{38} times per second. That's 1,000 trillion, trillion, trillion reactions in about the time it took you to read this sentence, which has been deliberately padded so it lasts 10 seconds – or thereabouts.

All that energy adds up. It keeps the Sun hot, allowing yet more reactions to take place. And the light slowly trickles out of the Sun, each photon of energy being absorbed and re-emitted again and again and again as it crashes into the densely packed particles in the Sun's core. In all, depending on your assumptions, it takes about 100,000 years for a photon to escape to the Sun's surface[17]. Then it's a short hop, skip and an eight-minute jump to reach the Earth where it is casually absorbed by your eyelid because you blinked at the wrong moment. Ah well.

Not all stars work the same way, but all are powered by nuclear fusion of some type or another. The nuclear furnace of our Sun is replicated across the Universe countless times, each with its own slight change to the formula. But the underlying physics is always the same – more mass in

---

[17] Though to be fair, it's not the same photon that was produced in the fusion reaction – which is a comment on how the energy packet we know as a photon has been split up and shifted around countless times in its path to the surface, not a suggestion that photons go through emotional and psychological changes in the ordeal.

the ingredients than in the products. Was it ever the other way around, well... let's not get ahead of ourselves.

You might be asking where is the proof for all of this? Well believe it or not you are bathing in it, right now. I say that with confidence, whether you are currently outside in the blazing sunlight or wrapped up in the dead of night – because the things you are bathing in are neutrinos.

Neutrinos are weird subatomic particles. Practically no mass, travel at practically the speed of light, and pass through almost anything – even the entirety of the Earth. Every one of those chains of reactions that produces energy in the Sun produces them, so that's countless numbers of super-fast, super-elusive particles streaming from the Sun every second. A small fraction, still numbering in the hundred trillions, passes straight through you. Thankfully, they aren't entirely unstoppable. Very,

very[18] rarely they happen to smash into the nucleus of an atom and a short flash of blue light is produced, known as Cherenkov radiation. Big detectors, like IceCube in Antarctica and Super-Kamiokande in Japan are constantly looking for these faint blue pulses – and their constant detection of these particles is just one way that we know what's going on in the centre of the Sun.

---

[18] very, very, very, very, **VERY.**

The Births of Stars

So, the core of the Sun is just a vast nuclear furnace, converting hydrogen into helium, resulting in all the lovely energy that lights our sky, powers plants and melts your ice cream just that bit too fast in those three weeks of the year the UK gets warm. But while we may know what our Sun looks like now, the question becomes how did it get that way?

Despite claims to the contrary, space is not a vacuum. Or to be more accurate, it's not a perfect vacuum. Between the stars there are vast quantities of gas and dust made of the same stuff as pretty much everything else in the Universe – three-quarters hydrogen, one-quarter helium, and maybe a few other trace elements thrown in for good measure. While exactly how much gas there is depends on the galaxy in question, there are often vast stores of it hidden between the stars. It's just that it's so spread out that space looks practically empty. Look in a cube of space one centimetre on each side and you'll find somewhere between one and one million atoms or molecules, depending on where you are in the galaxy. Do the same with the air of the Earth at sea level and you'll find 30 quintillion (30,000,000,000,000,000,000).

This sparce space gas, called the **interstellar medium** is what can eventually become stars, if given the chance. It isn't entirely smoothly distributed around the galaxy. Instead, some parts are much more clumpy[19] than other bits. Known as **nebulae** from the Latin for 'clouds', it's the

---

[19] Another technical term.

densest of these clouds that can become new stars.

These nebulae are far from the super-heated balls of gas that stars are. The temperatures of stars are so high that the **electrons** that orbit the nuclei of atoms like hydrogen and helium are stripped off. With so much heat energy around, the electrons can nick the little that they need to break away from their parent atom. They float around in a soup of electrons and stripped nuclei known as a plasma.

These gas clouds, however, are cold. Not only well below freezing, they start to push all the way down towards **absolute zero**, the coldest temperature anything can ever be. Perhaps unsurprisingly then, the various atoms in the cloud can stay whole, with electrons orbiting their nuclei as electrons tend to do. In fact, many of these atoms can start to stick together into more complex molecules like molecular hydrogen (H_2), carbon monoxide (CO) and even alcohol[20]! Hence another of their names, **molecular gas clouds**.

These clouds are vast, ranging in size from a few lightyears to several hundreds of lightyears across. Each can have thousands or even millions of times the mass of our Sun bound up in the gases inside them. Plenty of building material for new stars.

But how exactly do these vast clouds go from being cold and diffuse to hot and dense? Well, how else? Gravity.

While our Sun may not be powered by its own collapse now, it got to where it is because at one time it was collapsing. Some grand molecular gas cloud from about 4.5 billion years ago began to collapse under its own weight. But why it did so is the cause of some debate.

---

[20] Before anyone gets any ideas for a cosmic still, it's mostly methanol, rather than ethanol – so definitely not for drinking.

While gravity is a force that constantly tries to bring everything to some central point, other forces will try and stop it. This outwards pressure comes from the fact that atoms bounce. If you have a gas with lots of particles constantly bouncing off one another, this causes the phenomenon that we experience as gas pressure. Crush a gas down, either by squeezing the gas into a smaller volume or by putting more atoms of gas in the same volume, and you decrease the distance between each atom, making collisions more frequent and so raising the pressure. This is why it gets harder and harder to fill a balloon after each new puff of air goes in. Similarly, if you raise the temperature of a gas, the atoms in it move around faster, collisions are more common and the gas pressure goes up.

In a vast molecular gas cloud, the temperature and density are both extremely low, but so too is the gravity.

Yes, it has many thousands of times the mass of the Sun in it, but all of that mass is spread over a vast area, and gravity drops in strength greatly with distance. The result is that many molecular gas clouds are stable – they aren't collapsing… yet.

So what happened to our own cloud?

In order for a cloud to collapse, it has to be dense enough that its own gravity can overcome the gas pressure pushing outward. In other words, it might need a bit of a squeeze. Often this happens in one of two ways.

The first comes from the understanding that things in space move – even really big gas clouds. And because not everything is moving the same way at the same speed, eventually collisions are going to happen. Unlike when solid objects collide, collisions of gas clouds are not particularly abrupt. Instead they sort of pass through one another, but thanks to gas pressure they aren't completely unaffected. The front of each gas cloud slows down when it encounters the other causing something of a pile-up – a gas traffic jam as the rear of the cloud presses into the slower front. This is known as a shock, and means the density of the clouds at the intersection between them may just go over that critical density required to collapse under its own weight.

This can happen even if the clouds in question aren't originally from the same galaxy! Galaxies are constantly moving around too, and collisions are remarkably common. Our own Milky Way galaxy is on a collision course with the Andromeda galaxy and is expected to collide in about 4 to 5 billion years at about 4 o'clock on a Tuesday[21]. Galaxies are so diffuse that the chance that any actual collisions between stars or planets will occur is

---

[21] Which Tuesday isn't quite clear but personally I'm hoping for early spring.

practically zero. But what almost certainly will happen is that both galaxies (or more accurately the amalgamation galaxy, Milkdromeda[22]) will have a large burst in star formation as the gas clouds in each collide. Our own Sun may well have come from the collision of the Milky Way with a dwarf galaxy known as Sagittarius, which sweeps regularly through our galaxy and is estimated to have had one such collision only a hundred million years or so before our own Sun's birth. Coincidence? Possibly, yeah. But interesting all the same.

The second way to push a molecular gas cloud over the edge is a little bit more violent[23]. Some stars don't go quietly into the night, and although I'll leave the juicier details for when we get to my favourite section of this book, these explosive

---

[22] Worst. Galaxy. Name. Ever. But at least we have a few billion years to improve on it.

[23] Because colliding intergalactic gaseous big rigs into one another apparently isn't enough.

The surface of this ball of super-heated gas we call our Sun is constantly changing, producing plumes called prominences that are suspended in the atmosphere by the Sun's magnetic field.

NASA/GSFC/Solar Dynamics Observatory

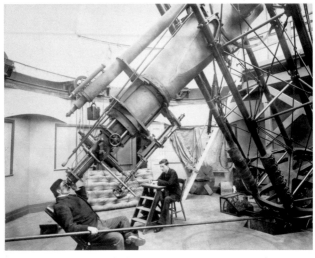

An astronomer and his assistant operating the Great Equatorial Telescope, a 28-inch refractor housed at the Royal Observatory Greenwich. The telescope is still in use today over a hundred years after this image was taken.
Royal Museums Greenwich

The dust and gas of the Witch Head Nebula (IC 2118) has a faint, ghostly appearance, illuminated by the nearby supergiant star Rigel.
NASA/STScI Digitized Sky Survey/Noel Carboni

Stars are usually just points of light to even the most powerful telescopes. But with very few we can see some detail on their surfaces or in their atmospheres, as the Hubble Space Telescope did with Betelgeuse in 1995.

Andrea Dupree (Harvard-Smithsonian CfA), Ronald Gilliland (STScI), NASA and ESA

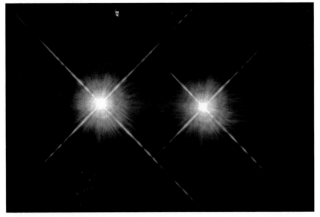

Alpha Centauri A (left) and Alpha Centauri B (right) taken by
the Hubble Space telescope. The Alpha Centauri group is the
closest star system to the Earth, with Proxima Centauri (not
pictured) being the closest of the three.
ESA/NASA

The Bubble Nebula (NGC 7635) in the constellation Cassiopeia being blown into space by a super-hot, massive star that is 45 times more massive than our sun.
NASA, ESA, and the Hubble Heritage Team (STScI/AURA)

The brightest star in the night sky, Sirius, taken from the ground. It is found in the constellation Canis Major and appears almost twice as bright as the next brightest star, Canopus.
Akira Fujii

There are more than 3,000 stars of various sizes in this dramatic image of the Orion Nebula taken by the Hubble Space Telescope.

NASA, ESA, M. Robberto (Space Telescope Science Institute/ ESA) and the Hubble Space Telescope Orion Treasury Project Team

supernovae aren't just the end of one star's life. They can be the beginning of an entire new generation. If this supernova goes off close enough to a molecular gas cloud, the blast wave can be just the push that's needed, creating the same sort of shock as a collision would. As we'll see later, the evidence for this method for triggering our own Sun's formation is compelling too, with much of what you see around you formed from elements that could only have come from the leftovers of a supernova.

So, was it a cosmic car crash or an interstellar explosion that helped form our Sun? We simply don't know, at least not yet. What happened afterwards, though, is at least a little clearer!

This cloud, now dense enough to collapse under its own weight, does just that. As parts of the cloud will be denser than other bits, the collapse won't be uniform. Denser chunks have stronger self-gravity and so will collapse faster, fragmenting the cloud into clumps. These clumps are themselves, well, clumpy, so they fragment even further. And so on. And so on. Each smaller and smaller fragment has the potential to go on to be the core of a star. This helps explain why this vast cloud of thousands of times the mass of our Sun doesn't just collapse into one absolutely vast, supermassive star.

But something does have to change at some point. If fragmentation went on forever, each clump would keep dividing and dividing until it was less than a puff of air from the lung of a gnat[24]. Clearly this would hardly be the stuff of vast celestial bodies.

The solution, it turns out, is a matter of insulation. As we discovered earlier when we briefly thought the Sun might be powering itself by slowly collapsing, when

---

[24] Which is even smaller than you might think given insects don't have lungs.

things fall they speed up. The interaction of this falling stuff with the other stuff around it slows it back down and converts the energy to heat and light, a bit like the underside of a spacecraft falling back to Earth. But we also learned earlier that when gases heat up, the pressure goes up, so it pushes back against gravity more strongly. As long as the cloud can bleed off the heat somehow it will continue to collapse, and while it's big and spread out, that's easy! Light can escape from it no problem, taking all that heat energy with it, so the collapsing cloud stays cool. A cool cloud means the gas pressure stays down and it can keep collapsing further and further, fragmenting as it goes.

But eventually the cloud gets dense enough that light can't easily get out. With the heat trapped in effectively a blanket, the cloud can't get rid of its heat as efficiently. It begins to heat up – not a lot, but enough to stop it from fragmenting uncontrollably anymore. Whatever mass a fragment is when it reaches this point is how much stuff it will have available to make its star – not that any of them will actually use it all. There's a lot of waste in star formation, which considering you are living on, and are made out of, that waste, is probably a very good thing.

Of course, if light can't get out, that makes seeing what's going on inside rather difficult. Thankfully, while visible light may be blocked, some types of light like infrared can still get through. Take a look at some of the most active star-forming regions in our (relatively) local neighbourhood like the Orion Nebula or the Eagle Nebula, and in visible light you'll see dark regions, impervious to your view. But take a look at those same regions in infrared light, and the sites of star formation come blazing through. It's like X-ray vision – just... not with X-rays.

The fragment that one day will become our Sun is much smaller now, less than half a lightyear (5 trillion kilometres or one-eighth the distance to our next closest star) across and roughly spherical. But it won't remain this way for long.

A dense centre forms with material falling slowly onto it, building up its mass. And it's now that something we've ignored comes into play. The original gas cloud was made up of vast numbers of particles of gas all travelling around in different directions and at different speeds, but the cloud itself is mostly travelling in one overall direction. It's like watching a flock of starlings just before they roost at night. Any individual bird is moving independently, and viewed on its own there seems to be little in the way of a pattern to its movement. But combine the movements of all of the birds together and the result is the flock shifts, slower than any one starling, but with far less apparent randomness involved.

But while the bulk motion of the cloud isn't particularly important to us[25], how it is spinning is. Add up all the motions of the particles going clockwise around the cloud and compare it to the sum of the motion anticlockwise and there will, inevitably, be a small imbalance. Though while the cloud is large it won't be noticeable, it is technically spinning. As it shrinks though, it begins to spin faster and faster, like an ice skater spinning on the spot and drawing their arms in.

But anyone who's been on a roundabout with a malfunctioning GPS will know that spinning around on the spot, in addition to being deeply disorientating, has an effect on how you move. As you travel around that disc of navigational misery, you'll be flung outwards. This is what has sometimes been referred to as the centrifugal force.

---

[25] After all, when in empty space, does it really matter if you are going up, down, left or right.

Technically, it's not a force at all – instead an example of something called inertia. Inertia is an object's resistance to changing its motion – whether a change in speed or a change in direction, both of which are known as an acceleration in physics. The more mass an object has, the more energy it takes to get it to accelerate by the same amount as something with less mass. It's basically a physical measure of matter's stubbornness towards change. As your car tries to take a turn around the roundabout, you, as an object with inertia, do your level best to keep going straight ahead. It is only by the grace of the friction between you and your seat, the restriction of your seatbelt, and, in some particularly sharp turns, the rigidity of the passenger-side window that you manage to make the turn with the car.

So what about our gas cloud? Well it has now reached the point where its spin is strong enough to allow parts of the cloud to effectively resist collapse. It's not resisting gravity as such – in fact, it's following the path that gravity has laid out for it. It's just that whereas before that path was mostly down and inwards to the centre of the cloud, now this bit of gas is moving sideways so fast that gravity can't make it fall downwards any further. Whirl a rock around on the end of a piece of string and you're asking for a night in A&E. But just like gravity, the string is always trying to pull the rock towards your hand. It just doesn't get there because it's moving sideways too fast. Right up until you slow it down. It's this same concept that allows the International Space Station (ISS) to orbit the Earth and keeps the Earth orbiting the Sun.

The thing is, orbits only work like that if the centre of the orbit is where the centre of gravity of the object it's orbiting is. The ISS moves in a big ring that is centred on the centre of the Earth, as that is the direction that gravity pulls. If someone tried to shift the ISS so that it orbited directly above the Arctic Circle at all times, the centre of its orbit would still be in the Earth but offset from the centre of mass by more than 5,000 kilometres. The result? The ISS would fall out of the sky – very, very rapidly.

So imagine our spherical cloud of gas as being like the Earth. It's spinning along its axis, as the Earth does along the line that connects the north and south poles. The stuff around the 'equator' of our cloud is supported by the fact it is now orbiting. But the stuff to the 'north' and 'south' is not. Like the misplaced ISS, it will try to fall. But unlike the ISS which would be a pile of wreckage on the Earth's surface, the cloud just collapses inwards. The top and bottom get crushed to join the stuff at the equator and you are left with a pancake – a Frisbee-shaped formation of gas and dust known as a **protoplanetary disc**. It is

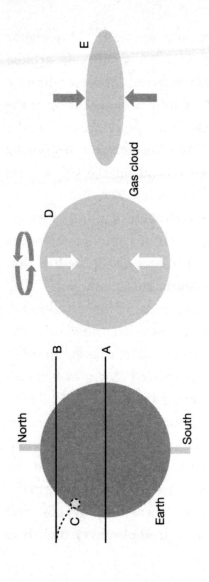

When the ISS orbits around the Earth so that it is centred on the centre of gravity of the Earth (orbit A, seen side on), it stays stably in orbit as its inertia keeps it from falling back to Earth. If, on the other hand, the ISS tried to orbit just the pole of the Earth (orbit B), it would fall back to Earth rapidly (path C). Similarly, while the equator of our spinning gas cloud (D) can support itself through its spin, the poles cannot, and the cloud flattens to a disc (E) over time.

from this that the young Sun will gather the last of its material, and what's left will eventually become the planets, moons, asteroids, and fluffy ducklings of the Solar System.

At the core, our fledgling Sun is drawing material from this disc, slowly growing denser and hotter. Hot enough now to have split up those molecules of hydrogen and to strip the electrons of the atoms that remain, the Sun is now a **protostar**. Briefly, the Sun begins nuclear fusion, but not of the simple hydrogen it does nowadays, but deuterium – heavy hydrogen, much rarer, but also much easier to fuse together into helium. This young Sun is violent and flaring, and flings jets of material off into deep space from its poles.

Finally, its core reaches the temperature needed to fuse simple hydrogen – 13 million degrees Celsius. A little bit of settling later and the Sun is all but the same as it is today. With vast quantities of heat being produced in its core, it can easily support itself against the incredible pull of its own gravity. Collapse stops, the last of the protoplanetary disc is either formed into objects or ejected into deep space. The Solar System is complete.

The whole process, from giant molecular gas cloud to bouncing baby hydrogen burner, is a mere 10 to 100 million years. Which may seem like a lot, but considering the lifespans of stars can be in the billions or even trillions, all of the stars in one molecular cloud are effectively born at the same time.

With your own unaided eyes you can see a cluster of stars that is still very much in its early years. The Pleiades, also known as the Seven Sisters because of how many you can typically see with your eyes, is a young **open cluster** of stars. At most 100 million years old, the stars within it are just as the Sun was shortly after birth. And just like the Pleiades, the Sun was in a vast cluster of stars – though not all were like the Sun. As each fragment of the original cloud was made from different quantities of stuff, some stars had more material available to grow from. Others had much less. So the Sun was surrounded by stellar siblings that were many times bigger and smaller than it.

But for some fragments, stars never quite formed at all. They would get close, but with too little matter available they could never crush themselves down enough to start fusion of hydrogen in their cores. These 'failed stars' are called **brown dwarfs**. They have somewhere between 13 and 80 times the mass of Jupiter, or less than 0.08 times the mass of the Sun.

And they don't really do very much. The larger ones might just be able to fuse Lithium or Deuterium for a while, but many can't even do that. With no reliable way to produce energy they mostly just slowly lose what little energy they gained during collapse into space. While the cores of these dwarfs can still be a respectable 3 million degrees Celsius, their surfaces can be surprisingly cold. A handful have been discovered that would be colder to the touch than human skin[26].

But enough of the failures – what about those that succeeded? What comes next?

---

[26] Presumably still attached to a living human, though NASA wasn't too specific.

The (Mostly Boring) Lives of Stars

Well, for many, not a whole lot for a really long time.

When each star begins the process of fusing hydrogen in their cores, they join something called the **main sequence,** and to explain what that is we are going to need... a graph.

Scientists, in general, love to graph things. Partly it's so they can show off a new plugin they found for their favourite programming language to all the other scientists. But mostly it's to look for patterns. Plot two quantities against one another, like height and shoe size, age and income, or US deaths by drowning in swimming pools and movies with Nicholas Cage[27], and you may start to notice some patterns.

Granted some of these patterns will be nonsense[28]. As the old adage goes, 'correlation does not imply causation', meaning just because two things look like they follow one another, does not mean they actually have anything to do with each other. However, sometimes that correlation will mean something and it's by finding these links that scientists can make progress in understanding the Universe[29]. Two such scientists in the early 1900s looking to find a pattern were Ejnar Hertzsprung and Henry Norris Russell.

---

[27] I promise you, I am not making this one up.
[28] I'm looking at you, Cage.
[29] And make some pretty swanky matplotlib graphics at the same time.

Astronomers had made many attempts in the past to categorise stars by what little information they had available to them. Antonia Maury, Williamina Fleming and the fantastically named Annie Jump Cannon, had worked on the spectra of stars, noticing how certain features were more or less prominent. This led to the Harvard Classification System, where each star is put into a class denoted by a letter – O, B, A, F, G, K and M. This has been remembered since the early 1900s via the mnemonic 'Oh Be A Fine Girl/Guy, Kiss Me' and by considerably less quaint phrases by university students since a short time later. The hot blue stars are type O, running down through the medium-hot yellow-white G-type stars, the type our own Sun is, to the cool red stars of K and M.

Hertzsprung, Russell and others took these classes and various other properties of stars, and plotted them against one another, looking for those important patterns. And, they found one. Take the luminosity or absolute brightness of a star, and plot it against its surface temperature or spectral class. Do that with a lot of stars and you get the Hertzsprung-Russell diagram – or H-R diagram for short.

If you've chosen a good mix of stars, and plotted enough of them, the most striking feature will be a line of stars running from high luminosity and high temperature down to low luminosity and low temperature. This slightly S-shaped curve of stars is known as, yep, the main sequence. It's where stars spend the majority of their lives. And it's here where you'll find our own Sun. Stars remain on the main sequence for the entirety of the time they are fusing hydrogen in their cores.

During this part of their lives, most stars are, well, boring. Yes, some stars are more active than others, with flares, and plumes and spots and giant coronal mass ejections.

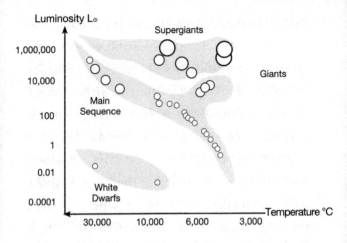

The Hertzsprung-Russell diagram shows different types of stars, from the middle-aged main sequence, through the older giants and supergiants, and onto the dead white dwarfs.

But all in all, they stay mostly where they are on the H-R diagram from the moment they start stably fusing hydrogen to the moment they run out of fuel – or at least until they run out for the first time. They stay mostly the same brightness, mostly the same temperature, mostly the same size for potentially a very, very long time. Don't misunderstand me though, stars being boring is a pretty good thing. If our Sun were a bit more 'exciting' chances are humans wouldn't be around to be thankful (or otherwise) for this fictitious Sun's general lack of monotony.

Exactly how long a star stays this way is dependent (almost[30]) entirely on how massive the star is, but possibly not in the way that you might think. Yes it's true that the bigger a star is the more hydrogen fuel it has available for fusion. This is like having a larger fuel tank in your car or a bigger battery on your phone. However, there's actually something else that's far more important.

Remember how raising the temperature of a gas made all the particles in that gas fly around faster, meaning more collisions and more gas pressure. Well more collisions also means more chances to fuse hydrogen nuclei together. Also, thanks to how

---

[30] We'll get to that.

thermodynamics works, each collision has a higher chance of success if the particles are travelling faster. In other words, the hotter the nuclear furnace is, the faster hydrogen is converted into helium. But, the faster hydrogen is converted into helium, the more energy is released. The more energy is released, the higher the temperature gets so the more collisions occur and so on and so on. Diminishing returns eventually halt this vicious thermal cycle, but the result is this. A small increase in the temperature in the core of a star, means a huge increase in the rate of fusion.

And which stars have the hottest cores? Well the ones with the largest masses! The compression due to gravity of these stars is intense, pushing the pressure and temperature through the roof. Because their cores are so active, they are producing by far the most light and hence are the most luminous. That heat also reaches up to the surface, making them the bluest, hottest stars on the main sequence. And having the most stuff in them, plus being puffed up by the intense heat inside them, makes them the biggest stars on the main sequence too. Finally, they are by far the rarest type of star out there. For every one of these most massive stars, there are a thousand stars like our own Sun.

The coldest stars on the other hand are the least massive. These tiny **red dwarf** stars had barely enough material to become stars in the first place, no more than a few times heavier than the largest brown dwarfs. These are the most common types of stars, with hundreds for every Sun-like star. As the name suggests, these are deep red in colour, owing to their low surface temperatures. With considerably colder cores, their lifetimes are many times longer even than the middle of the pack, the yellow dwarfs, of which our Sun is one.

But just how much longer? Our Sun is 4.5 billion years into its main sequence lifetime, which is a little less than halfway through its expected 10–12-billion-year span. The biggest, brightest and bluest stars can exhaust their immense supply in a matter of a few million years. This leads to a rather surprising result. Because star formation from cloud to star can take a few hundred million years, but the largest products of that formation die in only a fraction of that, one great way to look for active star formation is to look for the massive explosions that hail the ends of these stars lives. Stellar death as a marker of current stellar birth.

For red dwarfs things are slightly more complicated, however. The cores of medium and large stars stay mostly intact throughout their adult lifetimes. Although some stuff does rise up in the star and some stuff at the edges fall back down, for the most part what fuel there is in the core of the star is all the fuel it will have for its main sequence. Red dwarfs, however, are tiny and the whole star gets mixed thoroughly throughout its life. In other words, red dwarfs can use the hydrogen from the entire star. Add that to the super low rate of fusion and the longest living red dwarfs are expected to last well into the trillions, even tens of trillions of years. Given our entire Universe is only 13.8 billion years old, give or take a little bit, not one red dwarf star has yet had time to die – at least, not by running out of fuel, the star equivalent of death through old age.

As we'll see shortly, when stars end their time on the main sequence, they quite rapidly, at least by star standards, change their luminosity, size and temperature. This means a star that has recently ended its adult lifetime will quickly leave the main sequence on the H-R diagram. Using this, and the fact that stars leave the main sequence at different times depending on their mass, we can determine one of the most elusive properties of a star – its age.

Determining the age of a main sequence star is not easy. We can say it must be younger than the maximum length of time it can stay on the main sequence, but that's a bit like pointing at a random person and saying with confidence that they are less than 120 years old. It's almost certainly true, but it's not particularly helpful if you are putting candles on their birthday cake. The problem is that stars look almost exactly the same across their main sequence lifetimes, and even though they do evolve a little bit in this time, unless you know exactly where on the H-R diagram they started, it's rather tough to work out how long it's been since then.

But, if a cluster of stars all formed at the same time, then the brightest bluest stars will all die first, followed by the mid-sized white-yellow ones, and then finally the small red ones. Plot a H-R diagram for a real cluster of stars and chances are you won't have a full main sequence.

You'll have all of the lower part intact until you reach a sharp bend. That bend is where stars are coming off of the main sequence and heading into stellar old age. Those stars have just reached the age they need to be in order to leave their main adult lifetimes behind and become the star equivalent of an OAP.

Because all of the stars are the same age, this tells us how old the cluster is. The more intact and longer the main sequence appears, the younger the cluster. The shorter and less intact, the older. This method only works if you can be reasonably sure all of the stars are actually from the same cluster, and that that cluster had only one burst of star formation. Plus, it's useless if applied to single stars far away from any other group. But nonetheless, it's a good method for finding a very hard to establish piece of information about the stars in the sky.

The (Mostly Exciting) Deaths of Stars

When stars do eventually run out of hydrogen fuel in their cores, it is not necessarily the end just yet, though it's not too far off. Given the big differences between different types of stars we'll look at what happens with three groups – the low mass red dwarfs, the mid-mass Sun-like stars and the massive blue stars.

In all three cases, it all comes down to the interplay between two forces we are already very familiar with. Gravity pulling inwards, and pressure pushing back out. Throughout the star's formation gravity was winning, so the pre-star material shrunk. During its main sequence, pressure and gravity were about evenly matched so the star stayed basically the same. But once the fuel begins to run out, the nuclear engine begins to falter. With less energy being produced the temperature in the core decreases and the gas pressure goes with it. Gravity starts to win again.

For red dwarfs, this process is slow. It takes hundreds of billions of years for a red dwarf to collapse down from the main sequence, growing hotter and bluer as its core gets compressed under the vast weight of the star. Eventually it will crush itself down into a **white dwarf**. Gravity has compressed the object down so much that it can no longer be supported by gas pressure alone. Instead, all of the electrons in the star are being pressed up against one another. A law in quantum mechanics states that some particles, like electrons, cannot share the same space. Instead, in these white dwarfs, the electrons found in the star are confined to individual spaces, like the audience in a super-heated quantum mechanical cinema. As the seats fill up, strict adherence to the one-to-a-chair rule[31] means eventually they can't be crammed in any further. This resistance to further compression is called **degeneracy pressure**.

But even with this, the star is now incredibly dense. A single teaspoon of white dwarf material would weigh a tonne. Made almost entirely of helium, thanks to the red dwarf having used up almost all of its hydrogen, nothing much happens in these objects. There's no fusion in their core, no way of producing more energy. So they slowly cool down, taking hundreds of trillions of years more to become the hypothetical **black dwarf** – a dense, cool, boring lump of stuff. Unsurprisingly, no

---

[31] And absolutely no standing room allowed.

such object yet exists in our Universe to our knowledge.

Mid-mass stars, like our own Sun, have a rather more exciting journey to go on, even if the destination is a little similar. These are the stars that range from a little over half the mass of our Sun up to about eight times its mass. When they run out of fuel in their cores and gravity begins to win, the star begins to collapse in upon itself. But as we've seen before, crushing an object down just makes it hotter. Around a core made entirely of helium is a vast envelope of mostly hydrogen, the outer layers of the star. The inner layer of this envelope gets crushed against the core and heats up to the 13 million degrees Celsius needed to start hydrogen fusion again, this time in a shell surrounding the core.

This vast injection of new heat puffs up the outer layers of the star, greatly increasing the star's radius. The outer layer, now more distant from the hot core than ever, cools down considerably, turning from the normal yellow-white to a deep orange-red. A red giant is born. The star is far more luminous and far redder than before and gets pushed far to the top right of the H-R diagram.

When our own Sun enters this phase, it will expand far enough to swallow Mercury and Venus whole. Though our own Earth will likely escape that fate directly, it will be baked thanks to the entire surface of the Sun travelling a little over 100 million kilometres towards us as it expands. Consider taking a very off-world holiday before this happens, about 5 billion years from now – book now to avoid the rush.

For stars up to a couple time the mass of the Sun, the centre of the star is a bit like a covered helium white dwarf. Not doing anything, but slowly growing as the shell of burning hydrogen around it dumps helium onto it like ashes from a fire. It eventually gets big and hot enough for helium fusion to begin. The whole core suddenly flashes into life, producing a vast amount of energy. Exactly how much and how fast varies, but in some cases the core outshines the entire Milky Way galaxy for a second. All this energy, known as the **helium flash**, is absorbed by the star and is never seen on the surface. Slightly bigger stars, up to eight times the mass of the Sun also start fusion of helium in their cores, though rather less abruptly. Finally, with helium fusion done in the core, one more collapse and bounce causes the inner layer of the envelope to fuse hydrogen and helium in shells around the core again. The star gets bigger and more luminous than ever before. But its time has very much come.

During this last phase of its life, it has been rapidly blowing material out into deep space eventually losing almost all of its outer layers. What is left is a new type of white dwarf, one formed from carbon and oxygen, with only a thin atmosphere of hydrogen and helium around it. Thanks to red dwarfs lasting longer than our Universe has been around, and big stars being so comparatively rare, carbon-oxygen white dwarfs are by far the most common type of **stellar remnant** present in our galaxy. It will remain that way for some time to come, until the Universe is old enough for its vast supply of red dwarfs to finally begin to die.

Like helium white dwarfs, carbon-oxygen white dwarfs are held up by degeneracy pressure. They are about the size of the Earth, but have between 0.2 and 1.3 times the mass of the Sun shoved inside of them, making them up to 1 million times denser than the Sun. And just like helium white dwarfs, they are almost entirely boring. They produce no energy in their cores – they do almost nothing.

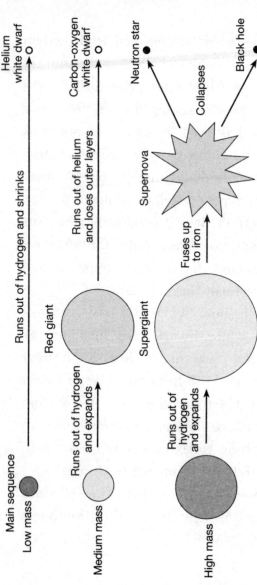

The evolution of a star, from main sequence death, is (almost) entirely dependent on mass. While big stars live fast and die young, small stars last for an extremely long time, and generally do very little with it!

90

I said almost entirely boring because there is one remarkable thing you can do with a white dwarf, and that is look for exoplanets: planets around other stars. But not the fully intact exoplanets around normal stars – oh no! The broken, lifeless, somewhat macabre remains of the exoplanets torn apart, either when their star expanded into a red giant or when they fell too close to this dense white dwarf remnant.

You see, if you take a spectrum of a white dwarf, you will see dark lines in it, like a barcode of the elements found in its atmosphere. For a carbon-oxygen white dwarf you would expect to see some carbon and oxygen, of course, and perhaps some hydrogen and helium from its thin atmosphere, but that's about it. But look at many white dwarfs and you'll find calcium, magnesium, silicon, iron, nickel and many more. These are traces of precisely the sorts of elements found in rocky planets, likely meaning we are seeing large chunks of rocky planets raining down on to these otherwise inactive dead stellar cores.

And now the bit we've all been waiting for[32] – the deaths of the most massive stars. After only a few tens of millions of years on the main sequence, our huge star, more than eight times the mass of the Sun, has exhausted the hydrogen in its core. Like more moderate stars it too goes through phases of fusion as gravity's brief victory against pressure results in parts of the star reaching that critical temperature needed for fusion of hydrogen to helium or helium on to carbon.

But it doesn't stop there. More phases occur as carbon fuses on up to neon, then to oxygen and silicon, all the way up to iron. Eventually the star becomes a vast onion-like structure, with deeper layers fusing

---

[32] Presuming that the reader is as much a fan of cosmic explosions as I am!

heavier and heavier elements. But iron is where the fusion ends. At least in the core of stars, iron is as far as we can go.

When the core is almost entirely composed of iron, fusion turns off extremely rapidly. The core rapidly cools and shrinks. The result is like smashing out the bottom layer on a house of cards. The whole star rapidly collapses down upon itself, rebounds on the extremely dense iron core and explodes. This vast explosion is called a **core-collapse supernova**. More energy comes out of the star in this brief explosion than is produced in the entirety of the Sun's lifetime. The explosion can easily outshine its host galaxy, being visible far across the Universe.

While the star itself had been able to do little more than build elements up to iron, this vast explosion with its huge heat and no need to support itself from collapse, causes the rapid creation of even heavier elements, including gold and silver. It's the fact that our own Earth has vast quantities of iron and nickel, plus the elements that can only be formed in extreme events like a supernova, that tell us that our solar system, and thus our Sun, are the product of a cloud of gas that has seen at least one supernova nearby. Possibly even one that triggered our Sun's formation in the first place.

What happens to the core of the star is yet again determined by mass. For stars up to about 30 times the mass of our Sun, the dense iron core is mostly crushed down, far beyond the realm of a white dwarf. Degeneracy pressure of electrons is not strong enough to resist the incredible forces present. The electrons and protons in the core are forced to combine with one another, forming neutrons, all bound together in a vast nucleus 20 kilometres across. Up to three times the mass of the Sun shoved inside a region half the width of the M25 motorway. Take the same teaspoon we stuck in our white dwarf from earlier, and this time it will have to support over a billion tonnes of material. Given it's also over half a million degrees Celsius, it's advisable not to use your mother's best silver for this one.

With such a huge mass contained in a tiny volume, this **neutron star's** gravity is exceptionally strong near its surface, 100 billion times stronger than on the surface of the Earth. Einstein's general relativity comes into play and things get weird. The gravity can bend the light coming off of the neutron star around its surface so that any observer can see both the front and back of it at the same time.

As with our spinning cloud of gas that formed the star in the first place, spinning faster and faster as it shrank, this tiny remnant has spun up from the relatively slow spin of an entire star to the entire neutron star spinning hundreds of times per second. Like the Earth, it has a magnetic field like a vast bar magnet. Unlike the Earth, it is billions to quadrillions of times more powerful.

Some neutron stars have jets of particles spewing from them pointing, in opposite directions out into space. As the neutron star spins, the jets, which aren't necessarily lined up with the axis of spin, draw huge circles on the sky. A distant observer that sits somewhere on that circle will see a pulse of light come from the direction of the remnant, repeating once per rotation. These **pulsars,** first noticed by then PhD student Jocelyn Bell Burnell in 1967, appeared so regular that one suggestion for their origin was extra-terrestrial intelligence – hence the first signal's rather tongue-in-cheek name LGM-1, the 'little green man'. Now, they can be used as precise clocks, testing theories of gravity across vast distances in space.

But what if the star that exploded was even bigger? Stars more massive than 30 of our Sun still end their lives in supernovae, but the collapse of the star onto its core is so great that there is no known force in the Universe that can stop it. The core collapses further and further, gravity overcoming every attempt to slow it down. The material of the core continues to fall in upon itself effectively forever, forming a point with no size, yet vast mass, many times that of our own Sun.

Or does it? To date, we've never been able to prove categorically that these **singularities** truly exist, but whether the core shrinks down to an infinitesimal point or just to something absurdly tiny, the view from the outside is the same. A **black hole**.

Contrary to popular belief, black holes are not cosmic vacuum cleaners. They have no more powerful gravity at a distance than any other object of the same mass. Replace our very own Sun with a black hole of one solar mass and little would change.

The planets would continue to orbit with almost no deviation in their path. Granted, all life on Earth would quickly die from lack of heat and light, and neighbouring alien astronomers might note that a very faint star had apparently disappeared overnight, starting an academic furore and a dramatically worded headline in the local newspaper. But in the grand scheme of things, nothing of any real consequence would have changed.

But get too close and things are vastly different. While a star or planet's mass is spread out, preventing its gravity from becoming too strong in any place, a black hole's mass is concentrated. Its gravity close to the centre is so strong that nothing, not even light itself, can escape if it gets too close. The point of no return is the **event horizon**, passing beyond which ensures that your future is to end up part of that compact central mass, whatever it may be.

And even before reaching this point, the force of gravity can break many objects apart. While standing on Earth, the force of gravity on your feet is very slightly stronger than the force on your head thanks to them being slightly closer to the Earth's core and its centre of mass. Thankfully on Earth this difference is incredibly minute so we don't notice it. Fall into a black hole feet first, however, and the difference between the force on your head and your feet can be vast. Your feet will begin to run[33] away from your body as you are rapidly stretched in a process known as spaghettification, presumably an altogether unpleasant way to go. As your body is stretched and compressed, and then begins to orbit the black hole before it finally falls in, it will begin to glow. This **accretion** of material by a black hole allows it to grow slowly over time. But

---

[33] This one was actually unintentional.

while it feeds, the glow from the disc of material around it can paradoxically mean that black holes, the ultimate dark objects, can shine more brightly than most stars. The glow of X-rays they produce mark them as being very different objects to the other points of light in our sky.

Ultimately, stars all have the same kind of death. They run out of fuel and they collapse, whether it be into a hot glowing white dwarf, a hot glowing neutron star, or a dark and mysterious black hole – each waiting for the end of the Universe.

Or are they?

Rarely, a remnant can encounter another object in space, usually within their own solar system. White dwarfs can merge with one another, or steal material off of another star, perhaps crossing the limit of how much stuff can be packed into one. It could collapse down into a neutron star itself, or trigger fusion in the white dwarf that quickly disintegrates the entire object in a rather different type of explosion called a **Type 1a supernova.**

Neutron stars too can merge with one another. The explosions they produce, a type of **gamma-ray burst,** is thought to be one of the main ways to produce the heaviest elements in the periodic table. Even black holes can merge with one another, the gravity of each tearing through space, literally stretching and distorting spacetime. The ripples that flow outwards from these extreme events can be 'listened' to by observatories on Earth as **gravitational waves.**

At least for now, even the dead remnant of a star is not necessarily the end of the story.

Why Nothing in Science is Ever That Easy

It really isn't.

As ridiculous as it may sound, what I have described for you in these pages is just the surface of stellar physics. The processes that govern stars are far more complicated even than this. The science in this represents only an approximation of the truth. The reason? Astronomers need to simplify things!

Physics is an extraordinarily complicated topic, and if we tried to understand everything in one go, we would simply crack under the (gas) pressure and get nowhere. Instead, we propose models for how things work that make assumptions about how things work. We simplify, remove the more complicated science and put it to one side. Then we test those models, see if our assumptions hold (at least mostly) true, and refine things from there, adding complexity when we understand the basics, a bit like learning to drive.

So, what assumptions have we used to make our lives easier?

Well early on we discussed how stars can often be found in pairs, trios or more! Multiple stars, born in the same system, orbiting one another. While many of these sets are so widely separated that they have little effect on each other, some can be so close together that at one time or another they will be able to steal material from each other, bulking up one star while draining the other. Could this change the path of a star's life? Studies say definitely yes!

While the earliest stars were made from the pristine material that came from the Big Bang (hydrogen, helium and the tiniest bit of lithium), later stars like our Sun have an array of materials from across the **periodic table** in them. Could these heavier elements have an effect on the processes happening inside a star? Probably!

Some stars pulse, changing brightness in a regular repeating pattern as they swell and shrink. Do stars like this live different lives to more stable stars? Perhaps!

Stars spin! Could a rapidly rotating star shake up its interior? Maybe!

Stars collide! Will smashing two stars into one another have any effect?... OK this is a definite yes. But how?

Sometimes research can be a tiresome thing. It's not all 'Eureka!' moments and celebrating a new monumental discovery. It is often slow, incremental, building on the work of astronomers stretching back decades, centuries or longer. And likely the work modern astronomers complete today will be built upon for many years to come.

That said, the 20th century provided more technological advancements in astronomy than any that had preceded it, with the invention and deployment of bigger and better telescopes, even ones capable of looking at types of light well beyond that which our own eyes could see. We have even begun to explore out into space – just our own little corner for now, but who knows what the future will bring. The 21st century is shaping up to continue this rapid development and we can expect impressive things about stars, galaxies and the Universe in general to be uncovered in the coming years.

Even so, it is all but guaranteed that every discovery will reveal yet more questions to answer.

Some would call this frustrating.

Astronomers call it job security.

Glossary of Terms

Absolute zero – the temperature at which all motion of particles stops. By definition, it is the lowest temperature any object can ever reach as all thermal energy has been removed. It is at -273.15°C or 0 **kelvin**.

Accretion – the process of an object gathering material from its environment through gravity in order to grow – a bit like a snowball growing in size as it rolls down a snowy hill.

Aperture – the opening at the end of a telescope that gathers light. Designed to be far larger than the human eye to improve light-gathering power compared to unaided stargazing.

Barycentre – the centre of mass of two or more objects. The point around which two objects will orbit one another. It will be skewed towards the objects with the largest mass and will only be at the exact middle point between two objects if their masses are identical. For example, with the Sun and Earth the barycentre is deep within the Sun. For the Sun and Jupiter it lies just outside the Sun's surface.

Binary – in astronomy, a system of two orbiting objects of the same type – in this case two stars.

Blueshift – a compression of light that shifts the light's type towards higher energy, or bluer, light. In the case of the Doppler effect, shows an object is moving towards us or, equivalently, us towards them. See also **redshift**.

Black dwarf – the theoretical end point of a white dwarf that has cooled completely.

The Universe has not been around long enough to have produced any.

Black hole – the remnant left behind after the deaths of the most massive stars.

Brown dwarf – an object that formed like a star, but was never large or hot enough to begin fusion of hydrogen in its core. A failed star.

Core-collapse supernova – the explosion that occurs at the end of a massive star's life. Begins as a collapse of the star that then explodes outwards.

Coronagraph – a device used to block the light of the Sun or a star in order to observe faint objects or atmospheres hidden in their glare. Similar to blocking the Sun with your hand on a bright day to see more easily.

Coronal mass ejection – a vast expulsion of material from the surface of the Sun, ejected by its magnetic field. Can potentially be harmful to electronic systems like satellites and power grids if the ejection is pointed at Earth.

Degeneracy pressure – a form of pressure that supports white dwarfs and neutron stars from collapse under gravity.

Electromagnetic spectrum – the entire range of types of light. In order of increasing energy, rough groups include: radio, microwaves, infrared, visible, ultraviolet, X-rays, gamma-rays. Further sub-groups exist for use in certain branches of physics.

Electron – a sub-atomic particle that often orbits the nucleus of an atom. Extremely light and with a negative charge.

Event horizon – the surface of no-return that surrounds a **black hole**. Not even light can escape from falling into a black hole beyond this.

Exoplanet – a planet that orbits around a star other than the Sun.

Filter – in astronomy, a device that blocks out certain types of light and allows through only the ones of interest in the observation.

Gamma-ray burst – an extremely 'gamma-ray bright' explosion associated with the deaths of the most massive stars, or with the merger of two **neutron stars**.

Granule – the top of a cell of hot rising gas in the Sun, similar to a single bubble rising in boiling water.

Gravitational potential energy – the stored energy that comes from lifting an object in altitude away from an object with mass (like the Earth). The higher it is, the more energy is stored. Dropping the object releases this energy.

Gravitational wave – a ripple left behind in the fabric of space by violent events like explosions and mergers that can be detected by observatories like LIGO and Virgo.

Helium flash – the burst of energy that results from the beginning of helium fusion in dying stars. Not visible from outside the star.

Interstellar medium – the diffuse gas that is present in the space between stars.

Kelvin – a temperature scale named after physicist William Thomson, 1st Baron Kelvin. A change in temperature of 1 kelvin is identical to a change of 1 degree Celsius. The difference between it and the Celsius system is its zero point. On the Celsius scale it is the freezing point of water in typical Earth conditions. On the Kelvin scale, it is **absolute zero**.

Kinetic energy – the energy associated with movement. The faster an object moves, the more energy it has.

Luminosity – the total amount of light emitted by an object per second.

Main sequence – the main group on a Hertzsprung-Russell diagram. Also the phase of a star's life where it is fusing hydrogen in is core – a middle-aged star.

Molecular gas cloud – a vast cloud of mostly hydrogen gas that is cool enough that simple molecules can form inside it, like molecular hydrogen and carbon monoxide. Contains the material from which stars are formed.

Nebula – a compact, dense cloud of gas. Some nebulae form new stars.

Neutrino – an extremely light, near light-speed particle formed as a result of nuclear reactions like fusion. Passes through materials with ease and only very rarely interacts.

Neutron – a sub-atomic particle that is a component of the nucleus of an atom. Slightly lighter than a proton and with no electric charge.

Neutron star – the remnant left behind after the **supernovae** of some massive stars. Only 20 kilometres across and composed entirely of **neutrons**.

Open cluster – a group of young stars that formed together. Though currently clustered, they will slowly drift apart over time.

Parallax – the apparent shift of nearby objects against a more distant background when changing the position of the observer. Useful for finding distances in astronomy.

Periodic table – an arrangement of all of the discovered elements in the Universe, based on the make-up of their atoms.

Photon – an individual 'piece' of light. A single packet of energy. How much energy the photon has determines what type of light it belongs to.

Photosphere – the last visible surface of the Sun and the place from which most of the Sun's light is emitted into deep space. Layers of the Sun below this are opaque and cannot typically be seen.

Plage – coming from the French word for 'beach', a bright spot found clustered around sunspots. However, the plages are found in the layer above the surface where sunspots are found. The exact connection is not fully understood.

Plasma – a substance that has had a large fraction of the atoms within it stripped of their electrons. Examples include the gas within neon lights and plasma televisions, flames and the material that stars are made from. This often occurs due to the substance being at a high temperature although irradiation with high energy light or strong electric currents can achieve a similar result.

Proton – a sub-atomic particle that is a component of the nucleus of an atom. Relatively heavy as subatomic particles go and with a positive charge. The number of protons an atom has determines what element it is.

Protoplanetary disc – the disc of material around a forming star from which planets and other smaller solar system bodies can form.

Protostar – a very young, almost formed star. Sits at the centre of a protoplanetary disc.

Pulsar – a spinning **neutron star** with a beam of particles and light. When the beam points towards the Earth, the pulsar appears to briefly brighten, or pulse.

Radiometric dating – a method that determines the age of an object by comparing the abundances of radioactive elements, and the products of that radioactivity. For example, comparing uranium and lead abundances in rocks of the Earth has helped determine the Solar System is 4.5 billion years old.

Red dwarf – a small star on the main sequence. Its red colour indicates its low temperature.

Red giant – a medium-sized star that is almost at the end of its life. It has swollen in size following running out of hydrogen in its core.

Redshift – a stretching of light that shifts the light's type towards lower energy, or redder, light. In the case of the Doppler effect, shows an object is moving away from us, or, equivalently, us away from them.

Refracting telescope – a telescope that bends and focuses light using lenses, similar to how spectacles work, as opposed to reflecting telescope that use mirrors to achieve the same goal.

Singularity – the theoretical centre of a **black hole**. A point with mass, but zero size.

Solar luminosity – the amount of light emitted by the Sun every second – 4×10^{26} Watts.

Solar mass – the amount of mass contained within the Sun – 2×10^{30} kg or about 330,000 times the mass of the Earth.

Spectroscope – a device that splits out the colours of light, producing a spectrum of the object it is observing.

Spectrum – in astronomy, a sorting of types of light in order of increasing or decreasing energy. Creates a rainbow pattern in visible light. Can also refer to a measurement of the intensity of light coming from an object across that range of types of light. See also **electromagnetic spectrum** and **visible spectrum**.

Standard candle – an object with a known luminosity that can be used to gauge distances by how faint it appears to be.

Stellar remnant – the remains of a dead star. Can be a **white dwarf**, a **neutron star** or a **black hole**.

Sunspots – regions on the surface of the Sun that are slightly colder than the surrounding area, making them appear darker in most images.

Supernova – a set of classes of explosions, often though not always involving the death of a star. See also **core-collapse supernova** and **Type 1a supernova**.

Thermodynamics – the physics of materials, radiation, heat and energy, and their interaction.

Type 1a supernova – a supernova explosion produced when a white dwarf grows too big, either through the merger with another white dwarf or the **accretion** of material from another star.

Variable stars – stars that change their brightness on much shorter timescales than their total lifetimes, like a lightbulb attached to a constantly changing dimmer switch.

Visible spectrum – a spectrum of light that is exclusively within the colours of light our eyes can see. See also **spectrum** and **electromagnetic spectrum**.

White dwarf – the remnant after the deaths of small and medium-sized stars. About the size of the Earth and composed of either helium or carbon and oxygen.

Royal Observatory
Greenwich Illuminates

Planets
by Dr Emily Drabek-Maunder
978-1-906367-82-4

Space Exploration
by Dhara Patel
978-1-906367-76-3

Black Holes
by Dr Ed Bloomer
978-1-906367-85-5

The Sun
by Brendan Owens
978-1-906367-86-2